Math Behind the Art

Math Behind the Art
Looking, Wondering, and Discovering Geometry

Text and artwork © 2026 Enrique Ortiz. All rights reserved.

No part of this book may be reproduced or transmitted in any form without written permission from the publisher.

ISBN: 979-8-9955799-0-8

First edition

Published by Enrique Ortiz Publishing
Oviedo, Florida

There are many ways to see and wonder.

**Math and art begin with careful looking
and thoughtful wondering.**

You can open this book anywhere
and use it any way you like.

Shapes + Colors = Me

WHAT DO YOU SEE?

- Shapes that repeat?

- Colors that sit next to their friends?

- A shape that stands out?

WHAT DO YOU WONDER?

- If one shape moved, would the picture feel different?

- How many ways could these shapes fit together?

Math Behind the Art

Artists group shapes to create balance. When shapes move without changing size, we notice position and pattern.

I'm a Star

I'm a Star

WHAT DO YOU SEE?

- Points meeting in the middle?
- Lines that look equal?

WHAT DO YOU WONDER?

- How many points are there—could we count in more than one way?
- Would it look the same if it turned? What about upside down?

Math Behind the Art

Stars show radial symmetry. Turning without changing shape is a rotation.

Paragon Bird

WHAT DO YOU SEE?

- A bird's outline filled with shapes?

- Spaces that feel like feathers?

WHAT DO YOU WONDER?

- How many shapes make the wing?

- Could a different outline change the feeling?

Math Behind the Art

A silhouette can be made from many smaller shapes.

This is composition—building a picture from parts.

Redshift Rotation

WHAT DO YOU SEE?

- Lines that seem to travel?

- A path your eyes want to follow?

WHAT DO YOU WONDER?

- Where might this path go?

- What happens if the picture flips?

Math Behind the Art

When shapes line up or tilt, they suggest direction.

Sliding a shape without turning is a translation.

Nested Balance

WHAT DO YOU SEE?

- Shapes inside other shapes?
- Equal spaces–or almost equal?

WHAT DO YOU WONDER?

- How many layers can we count?
- What happens if one layer grows?

Math Behind the Art

'Nested' means one shape lives inside another.

Artists play with size and spacing to create balance.

Modern Muse: Midnight Edition

WHAT DO YOU SEE?

• Tiny patterns hiding in bigger ones?

• Places that feel busy, and places that rest?

WHAT DO YOU WONDER?

• Where would you start drawing if you made this?

• What might you change to make it calmer—or wilder?

Math Behind the Art

Repeating and varying rules build structure:

repetition, contrast, and scale.

Portrait of a Vertical Dilemma

WHAT DO YOU SEE?

- Strong vertical lines?

- Shapes that almost tip–then don't?

WHAT DO YOU WONDER?

- What tiny change would make it fall–or feel safer?

- Which part does the most 'holding'?

Math Behind the Art

Designers use alignment to suggest stability.

Small shifts in angle or weight can change the balance.

Going Parabolic

WHAT DO YOU SEE?

- Arcs that open like smiles?

- Curves that get steeper?

WHAT DO YOU WONDER?

- Where would a rolling ball go?

- What changes if the curve opens the other way?

Math Behind the Art

A parabola is a special curve.

Sliding, turning, or enlarging the same curve are transformations.

Mocha Mousse Cylinder - Optical Illusion

WHAT DO YOU SEE?

- A picture that looks 3-D?

- Edges that fool your eyes?

WHAT DO YOU WONDER?

- Which clues make your brain say 'cylinder'?

- What happens if we cover part of the picture?

Math Behind the Art

Artists use light, shadow, and contour to suggest depth.
Your brain turns 2-D clues into a 3-D guess.

Necker

WHAT DO YOU SEE?

- A shape that can face two ways?

- A box that flips if you stare?

WHAT DO YOU WONDER?

- Can you make it flip on purpose?

- Which corner feels closest now?

Math Behind the Art

Some drawings are ambiguous—they allow more than one correct view.

V-Formation

WHAT DO YOU SEE?

- Shapes traveling like a team?

- A big 'V' made of smaller pieces?

WHAT DO YOU WONDER?

- What angle makes the 'V' feel just right?

- How many shapes keep the pattern going?

Math Behind the Art

Arranging shapes at an angle creates direction and unity. Patterns can be made from parts that keep their size and spacing.

25
9
16

Pythagorean World

WHAT DO YOU SEE?

- A right-angle corner?

- Squares that match the triangle's sides?

WHAT DO YOU WONDER?

- Could other numbers work too?

- Where else might we see these shapes?

Math Behind the Art

In a right triangle, the squares on the shorter sides can add up to the square on the longest side.

TRY IT

Pick one artwork.

Look

Look quietly.
Notice shapes, colors, lines, and spaces.

Wonder

What do you see more than once?
What might happen if one shape moved?
You can talk, point, or draw what you notice.

Look again

Did you see something new after looking longer?
You don't need to answer every question.
Looking and wondering are already doing math.

GLOSSARY

Translation – sliding a shape

Rotation – turning a shape

Symmetry – matching halves

About the Author

Enrique Ortiz is a mathematics educator and artist who explores geometry through abstraction. He teaches mathematics education at the University of Central Florida and creates artwork that invites viewers to look closely, ask questions, and discover patterns in unexpected places.

For teachers, parents and caregivers:
Access the free companion guide here:

This book grew out of classroom conversations and studio experiments—moments when noticing and wondering led to deeper mathematical thinking.

You don't need to know the math before you begin.

Start by looking. Then wonder.

Math Behind the Art is the first book in a series that explores geometry through art and inquiry.